YOU'RE ON MUTE

$10^{1/2}$ strategies for success in a Zooming world

PHIL BEDFORD

INDIA • SINGAPORE • MALAYSIA

Notion Press

No.8, 3rd Cross Street,
CIT Colony, Mylapore,
Chennai, Tamil Nadu – 600004

First Published by Notion Press 2021
Copyright © Phil Bedford 2021
All Rights Reserved.

ISBN 978-1-63781-695-0

This book has been published with all efforts taken to make the material error-free after the consent of the author. However, the author and the publisher do not assume and hereby disclaim any liability to any party for any loss, damage, or disruption caused by errors or omissions, whether such errors or omissions result from negligence, accident, or any other cause.

While every effort has been made to avoid any mistake or omission, this publication is being sold on the condition and understanding that neither the author nor the publishers or printers would be liable in any manner to any person by reason of any mistake or omission in this publication or for any action taken or omitted to be taken or advice rendered or accepted on the basis of this work. For any defect in printing or binding the publishers will be liable only to replace the defective copy by another copy of this work then available.

10^{1/2} Strategies for success in a Zooming World

Is your business going video first?

Are you collaborating, selling or networking more often online, rather than in-person?

Do you need to learn how to master your video conferencing experiences?

Then you're in the right place.

Few companies in any industry will ever experience growth at the same magnitude of Zoom video communications.

Founder Eric Yuan, and Zoom have achieved some truly phenomenal success, with a customer growth of around 354%! In 2020, Zoom became one of the leading tools for video conferencing in the world, and a must-have solution for companies of all sizes.

In a landscape where face-to-face interactions aren't always possible; Zoom allows individuals to virtually interact with their coworkers and connect over pressing projects.

Even individual consumers started to embrace Zoom in 2020, discovering the benefits of quick and easy video meetings for connecting with family and friends.

In this quick and simple book, we're going to cover **everything** you need to know about using Zoom Meetings and getting the most out of your virtual conversations. Crucially, although we refer to Zoom regularly and the specific features of the service in this book, you should be able to apply many of these tips to any video conferencing service.

Read on for our $10^{1/2}$ top Zoom Meeting tips, a complete overview of the Zoom application, and all of the guidance you need to discover Zoom success.

Contents

Chapter 1	An Introduction: What is Zoom?	7
Chapter 2	Tip 1 for Zoom Meetings: Video is Best	12
Chapter 3	Tip 2 for Zoom Meetings: Master Timing	17
Chapter 4	Tip 3 for Zoom Meetings: Prime the Technology	22
Chapter 5	Tip 4 for Zoom Meetings: Dress to Impress	26
Chapter 6	Tip 5 for Zoom Meetings: Be Interested	31
Chapter 7	Tip 6 for Zoom Meetings: Back Up Your Brand	36
Chapter 8	Tip 7: Design Clean Presentations	41
Chapter 9	Tip 8: Stand Out	47
Chapter 10	Tip 9: Master Your Background	51
Chapter 11	Tip 10: Follow up On Meetings	57
Chapter 12	Tip $10^{1/2}$: Go Back Again	63

CHAPTER 1

An Introduction: What is Zoom?

What is Zoom

Before you learn how to make the most of Zoom Meetings, you need to understand what it is, and how it works. Zoom is a cloud-based conferencing tool designed to simplify video conversations. With Zoom, you can virtually meet with other people.

Although it's possible to connect through Zoom's audio only, or even Zoom Chat, Zoom Meetings is all about video. In years gone by, companies had to use complicated hardware and software setups to connect via video. Only the biggest businesses with the largest budget could afford this tech.

Zoom changed all that by giving people a way to communicate over video with just one click. In 2019, over half of all Fortune 500 companies said they used Zoom for meetings. In 2020, the number of Zoom users skyrocketed, prompted by the rise of the work-from-home landscape, and the COVID-19 pandemic. Now that countless organizations have discovered, first-hand just how valuable Zoom can be, everyone wants to know how to make the most of this incredible tool.

The Features of Zoom: From Meetings to Rooms

When most people talk about Zoom, they'll be referencing the Zoom Meetings experience – the all-in-one video solution for connecting (almost) face-to-face. However, Zoom is actually an entire ecosystem of products, including:

- **Zoom Webinars:** One-to-many broadcasting functionality for training, education, and announcement purposes.

- **Zoom Rooms:** Zoom hardware solutions intended to transform standard office spaces into collaboration hubs. These rooms may include video cameras, microphones, speakers, and even virtual whiteboards.

- **Zoom Chat:** Zoom Chat is a lesser-known feature of Zoom offering easy messaging, similar in style to Slack or Microsoft Teams chat.

- **Zoom Phone:** This is a PBX replacement option for companies in search of business telephony through the cloud.

In this book, we're focusing on **Zoom Meetings.** Zoom Meetings is the core solution for Zoom, offering quick and easy video conversations in an instant. The platform allows for remote and co-located employees to communicate seamlessly. You can even conduct interviews with remote candidates over a Zoom Meeting.

To join a Zoom Meeting, you'll need a Zoom account (we'll discuss these in a minute), a webcam or video hardware solution, and a microphone. A set of speakers or headphones will allow you to hear the other person talking. You can download the Zoom Meetings app for your smartphone or computer or connect to a conversation via your web browser.

Although the features of Zoom Meetings are constantly evolving, at the time of writing, they include:

- **Virtual backgrounds:** Remove distractions from your background with a virtual replacement. Set a new background to hide clutter, or just spice up your conversations. Options include everything from offices to beaches.

- **Calendar integration:** Integrate Zoom with your Outlook calendar. Schedule your meetings as usual, then just click "Make it a Zoom Meeting" to add contacts to your invite.

- **Waiting rooms:** Make sure that your contacts have a place to chat safely while they wait for the meetings to start. You can admit one participant at a time, or everyone at once.

- **Breakout rooms:** Create dozens of breakout rooms where people in your community can chat amongst themselves and share information.

- **Multi-share:** Share multiple screens with coworkers at the same time. You can even decide which screen you want to see on your display.

- **Personal meeting rooms:** Create a personal meeting room ID that's completely unique to you. Adjust your settings according to your personal preferences.

- **Touch-up:** Don't let a rough night ruin your meeting. Touch up makes up for bad lighting and imperfections by smoothing your appearance.

Ways to Use Zoom: Web, PC, Phone, and More

There are tons of ways to use Zoom. The desktop app is available for MacOS or Windows. You can use this to launch a meeting from your computer whenever someone invites you with a meeting link. You can also create Zoom meetings too. You don't have to sign in to launch a meeting, but you can access extra features if you use your Zoom account.

Desktop users can:

- Start or join meetings
- Mute/unmute microphones
- Start/stop video
- Invite new people to a meeting
- Change your screen name
- Chat in-meeting
- Start a cloud or local recording
- Create polls
- Broadcast on Facebook Live

The desktop app offers slightly more functionality than the mobile app. With the mobile app, you can only record to the cloud, and there's no polling or Facebook Live feature.

If you prefer to avoid the desktop download, you can use Zoom as an Outlook plugin. This feature works within your Microsoft Outlook email client. Once installed, the service drops a button for Zoom Meetings onto your Outlook toolbar. With this button, you can start and schedule meetings with your contacts with just one click.

Meetings in Your Browser

Another option for Zoom Meetings is to start a conversation in your favorite browser. There's a Zoom Chrome extension, and a Zoom Firefox add-on that lets you schedule a meeting via Google calendar. A simple click on the Zoom button is enough to launch a meeting here. You can also send invitations automatically to your Google Calendar contacts.

Using Zoom in your browser is tricky if you don't have the app too – but it is possible. You can use a web client link that looks like this: zoom.us/wc/join/meeting-id.

There's also a browser extension that makes browser-based meetings easier. This is ideal if you're on a secure laptop or computer that might not permit the installation of apps. The extension is available through Firefox and Chrome – but it's not officially from Zoom, so be careful.

Chapter 2

Tip 1 for Zoom Meetings: Video is Best

Video conferencing has been an option for businesses for quite some time now. However, it's only recently that video made the transition from being a "nice to have" feature to one that's absolutely essential. 96% of people in an Owl Labs study say that video conferencing is essential to team communication – particularly with remote team members.

When the pandemic arrived in 2020, the demand for video conferencing grew. Companies needed a way to connect human beings in a world where face-to-face interactions weren't possible.

Psychological study tells us that human beings need face-to-face interaction to feel comfortable. While we can share information with nothing but an email address or an instant messaging app, video mimics our interactions in real life.

Without video, a significant portion of the conversation disappears. The context offered by facial expression, body language, gestures, and eye contact disappears when we're not together.

Video isn't just about more meaningful conversations either. With video conferencing, companies can:

- **Save money on travel**: There's no need to spend a fortune on plane tickets and travel if you can connect with your colleagues, partners, and contractors via video. Video gets us in the same virtual environment as our contacts quickly and cheaply.

- **Improve productivity:** How often do people on the other end of an audio conference spend their time answering emails and doing other work? When you're on-watch in a video call, you're more likely to focus on the task at hand.

- **Add context:** People in videos can show images and whiteboards, or just demonstrate their current emotional state with facial expression. Video adds context to a conversation that you can't get through audio alone.

Make Video Work for You

If you're going to be using a tool like Zoom Meetings in your business operations, don't be the guy that sticks with audio only. If you absolutely have no way to access a web cam or camera, then use a picture of you at least. This will give you some presence in the conversation. People with only their name available in a meeting are the least visible in any call. You would't attend a meeting with a bag on your head in the real world.

Video conferencing in the age of the remote and distance worker are all about **presence.** If you can't be present with video, an image will do. However, video boosts your chances of people remembering what you have to say.

To make the most of your Zoom video:

- **Change your background:** Embarrassed by the messy office behind you, or the kids that are constantly running back and forth out of shot? Change things up. Virtual backgrounds on Zoom allow you to situate yourself wherever you choose. Go into **Settings > Virtual background** and select the image you want for your background. You can put yourself in a forest, on a beach – or just in a neater office.

- **Turn on the beauty filter:** You know that you should wash your face and brush your hair before a Zoom meeting. Still, sometimes no matter what you do, you don't feel that you look your best. Zoom's Touch Up feature can help to give you more confidence in your conversation. Click on the **Video Settings** arrow next to Start Video and check the box that says **Touch up my Appearance.** You'll appear more well-rested and less blemished.

- **Adjust your camera settings:** If your image issue isn't with you – but with your camera, sort it out before you start the meeting. You can test your camera on Zoom before you join a meeting by clicking on your Profile picture, then Settings. Click the Video tab to see your image and the camera you have selected. You can adjust things like brightness and contrast here to improve image quality.

- **Use the name bar:** If people want to connect with you after a big group call they need to be able to find you on platforms like LinkedIn. Using just your first name in Zoom makes it much tougher to track you down later.

- **Get face positioning right:** Make sure you're sitting in just the right place, and have the camera focused on your face. There should be a little gap above your head, so that you take up most of the video stream box.

With extra features to make you feel more confident, you don't have to feel self-conscious about going on video with Zoom. Remember, these face-to-face interactions will improve your bond with the people you're meeting.

Remember to Set Up Your Space

Video is all about the visuals, so ensuring that you have a space that's set up for success is crucial. Investing in an SLR that gives you a fantastic HD video is a good first step but think about what's going on around you too. If you have partners, children, or pets that might be running around in the background, it may be a good idea to move to another room.

Having a separate space for your video conferences will help to reduce distractions. However, it's also a good way to reduce the blur between your work and life. When working from home, it's challenging to find the line between professional and personal.

After moving to the right room, make sure that you place your camera at eye-level, so you can make it feel as though you're making eye-contact with your colleagues. If there's a lot of mess in the background, you can blur it with the blurring settings on Zoom or the correct camera lens. A virtual background may be a little more appealing than blurring for some, however.

Other points to remember when setting up your space include:

- **Lighting:** Make sure that your face is well lit. Natural lighting often works best, and side lighting might be better than light head-on. Backlighting can often make it difficult to see, and lighting that comes from under your face will make you look more shadowy.

- **Clutter:** Clean the area around you to ensure that there's not many extra components battling for your audience's attention. This might not be too much of an issue if you're using blurring or a virtual background, but it's worth making an effort.

- **Clean up:** Aside from just cleaning up your space, remember to clean yourself too. You need a fresh face, and brushed hair, so it doesn't look as though you've just rolled out of bed. Ensuring that you wear the right work attire is important too. Just because you're meeting on Zoom doesn't mean that you can dress in your PJs. What if you forget what you're wearing mid-conversation and stand up to grab something? Embarrassments like that are more common on Zoom than you'd think.

Chapter 3

Tip 2 for Zoom Meetings: Master Timing

Timing is an important aspect of any professional life. Whether you're turning up at the office, or you're interacting with clients, you must be punctual. The same applies for any Zoom video conferences you're going to be a part of.

It's worth getting into the mindset that your Zoom meetings are just like any other meetings you're going to be a part of. You're expected to be there on time, not fiddling around with camera settings while everyone is waiting for you to log in. The best option to ensure that you're not lagging behind your team is to start setting up your meeting a few minutes early.

Load Zoom up ahead of schedule, and spend a few minutes ensuring that everything is set up just the way that you want it. This could mean that you take some time to fiddle around with the beauty settings on Zoom and remove any bags or dark circles you have under your eyes. It could also mean that you conduct a microphone check and ensure your voice is coming through clean and clear.

Ultimately, turning up late has the same impact in the virtual world that it has in the real world – you're going to come across as disorganized, or just disrespectful. Let's look at some more ways that you can improve the timing in your Zoom meetings.

Prepping for a Well-Timed Meeting

Being on time is just the first step, you also need to ensure that you're properly organized. Start well before the actual meeting takes place when you're planning a conversation with your peers. Remember that in today's digital world, there's a good chance that you'll be conversing with people on the other side of the world.

Before you suggest a meeting, check the time zones for everyone involved, and ensure that your contacts are going to be comfortable meeting at your required time. It's no good choosing a time that's going to be unsuitable for a handful of your attendees, as this just means that you have to reschedule. Suggest times well in advance so that your colleagues can request adjustments if necessary.

After everyone has agreed to a specific time, you can begin planning your upcoming meeting. An agenda is a valuable component of any professional meeting that many business leaders make the mistake of overlooking. However, agendas ensure that everyone stays on the right track, while also providing your attendees with the hints they need to make sure they're ready for the meeting.

Sending an agenda alongside the invite to the meeting will inform your contacts of whether they need to bring certain information with them to the discussion. This ensure that you can make the most of the time you have in the meeting, rather than waiting for people to rush around and find extra resources during the conversation.

Getting Familiar with the Tech

Zoom is one of the most popular video conferencing solutions in the world today – in part because it's so easy to use. However, it's still a good idea to make sure that you're comfortable with the technology before you dive in.

Start by checking that your camera and audio equipment work as they should. Your camera needs to be set up in a way that makes it appear as though you're sharing eye contact with your colleagues. Make sure that this is the case before you join the meeting, and make any adjustments to brightness, contrast, and other image elements before you log in.

Be acquainted with video conferencing features on Zoom that you're going to be using too. For instance, if you're going to be using mute to eliminate background noise, it's important that you have this feature set up in advance. If you're co-hosting a call, ensure that you know how to set up control for both you and your partner.

If you want to co-host a conversation, then you need to establish this feature in the Zoom Meeting Settings section first. Look for the Meeting tab, then click on **co-host.** When you start the meeting, wait for your new co-host to join, and add the person by clicking on the dots that appear when you hover over their video screen.

Another way to enable co-hosting is to go to the Participants window and click on **Manage Participants.** From there, hover over the name of the co-host you want, and select **More,** to find the co-host option. You will need a professional version of Zoom to access this feature.

Other features to get familiar with include:

- **Recording:** Do you need to record the video? If so, click on the three dot menu icon and tap on **Record.** This will ensure that you can share the conversation with people who weren't able to be here later.

- **Breakout rooms:** You can assign users to breakout rooms before your meeting begins. Just sign into your Zoom web portal, then click on **Meetings** and **Schedule a Meeting,** in the Meeting Options section, click on the **Breakout Room pre-assign** solution, and click on Create Rooms.

- **Audio transcription:** To avoid re-typing all of your meeting notes later, use audio transcription. You can easily search for information in a discussion this way. To enable audio transcription, sign into your Zoom Web Portal, then click on **Account Management,** go to the Cloud Recording option, and click on **Recording** to verify it's enabled.

Remember to Turn Up in Style

You don't have to make a grand entrance when you start a Zoom meeting, but you should make sure that you appear as professional as possible. Remember, this is still a professional conversation, and you need to treat it as such. That means no sweatpants and sweatshirts, and no PJs. Make sure your face is clean and your hair is neat.

Avoid wearing any patterns that might be distracting on-camera. Remember that wearing a bright white shirt may also get your camera to start auto-adjusting the brightness so it's harder to see your face. Soft colors are generally a better choice and solid colour jackets.

In general, you should dress as you normally would for a business meeting. If nothing else, this will help to get you in the right frame of mind. Just be aware of the fact that you're also on video. If you're using a greenscreen to create a more advanced virtual background experience, then you'll obviously want to avoid green clothing that might blend with this.

Remember, changing your clothes from standard home wear to business attire is also an excellent way to separate your home and work life. This is crucial if you're working from home.

CHAPTER 4

Tip 3 for Zoom Meetings: Prime the Technology

We've already mentioned that having a good Zoom meeting means ensuring you know how to use your technology effectively. However, there's more to this than simply knowing how to use the basic features of Zoom. You'll also need to check your internet connection, and make sure that your video and audio components are working as they should.

Start by making sure that you're comfortable with scheduling a meeting on Zoom. There are various ways to do this. You'll need to be a "host" to set up a meeting. Options include:

- Scheduling your meeting from the mobile or desktop app
- Scheduling from the Zoom web portal
- Scheduling a meeting for someone else

To launch a meeting from the Zoom desktop or web client, open the service and sign in. You can then click on the icon named **Schedule** and you'll see a calendar window. When you pick a date for your meeting, enter information about when it's going to start, how long it's going to last, and the time zone. If you want to set up a recurring meeting, you can click **recurring meeting** to do this.

Once your meeting is ready, you can invite participants either using a randomly generated unique meeting ID, or a personal meeting ID.

You'll also have the option to adjust:

- **Security:** Enter a meeting passcode, and your participants will be required to enter this before they can join the conversation.

- **Waiting room:** If you want to make sure that your employees have a place to chat while they wait for a meeting to start, use this service.

- **Video:** You can choose whether you want your video to be on or off when the meeting begins. You can also decide whether to turn video on and off for participants.

- **Audio:** Audio settings allow users to call in using computer audio, telephone, third-party audio, or something else entirely.

- **Dial in from:** If you want your contacts to be able to dial into a meeting, you can choose which countries they're able to dial in from.

If you have a calendar service set up for your employees, select this to add details too the meeting. Open the outlook desktop app and create an event for the meeting to get started. You'll see Outlook when using the windows client. You can also choose Google calendar, or connect to a third-party calendar.

Advanced options for your Zoom meeting include:

- **Join before host:** Allow your participants to join the meeting before you join

- **Mute participants on entry:** This ensures that no-one can speak before you

- **Authenticate:** This will restrict access so that only authenticated and signed-in users can join

- **Record:** Set your system to automatically record the meeting locally or to the cloud

- **Enable additional data centre regions:** For storing your information

- **Alternative hosts:** List the email addresses of other hosts for your meeting

- **Interpretation:** Enable language interpretation for your meetings

Click Save to finish setting up your meetings.

Check Your Wi-Fi, Microphone, and Camera

Once you know you're using Zoom correctly, it's important to check all the peripherals and extra technology you're going to be using, starting with your web connection. If you don't have a great connection, then your video is going to keep lagging out, which makes it much harder to have a clear conversation with your colleagues.

If possible, it might be worth using an ethernet cable to set up a wired connection for your conferences, this will reduce the risk of your connection dropping. It's also important to make sure that your Wi-Fi is secure, with a password protected link and advanced encryption.

Once your connection is properly established, check the quality of your microphone and camera. Remember that you want to make a good impression, so a high-quality camera with a clear image is crucial. It's also worth ensuring that your microphone can carry your voice well too. If your contacts are struggling to hear you, then you're sure to have some misunderstandings.

If necessary, upgrade your camera and microphone in advance, and make sure that they're properly set up to work with your Zoom call. It's also a good idea to make sure that you have your **mute** settings established so you can avoid distracting users when your colleagues are talking. When you're called to speak, you can hit the **space bar** on Zoom to mute and un-mute your meetings.

Remember, double check the quality of both your camera, and your microphone by running a test meeting with a friend before you begin any professional discussions.

Chapter 5

Tip 4 for Zoom Meetings: Dress to Impress

When Zoom and video meetings started to become more popular in recent years, an interesting trend emerged. Some employees, entrepreneurs, and freelance workers stopped making an effort with their appearance. They assumed that just because they were joining a meeting from home, they didn't have to dress in business attire.

Unfortunately, that can be a dangerous assumption. Whether you're joining a conversation in the real-world or over a Zoom feed, you **need** to have the right image. You wouldn't turn up to a conversation with a potential client or a coworker in a real-world office wearing your Pajamas. You wouldn't sit around eating your breakfast in the middle of a presentation.

When you're on Zoom, remember that you're still representing your business or professional brand. Don't let your image falter because you're too lazy to get dressed.

Making the Right Impression with a Great Image

Start by considering your surroundings. How will they affect the way you look? Lights and angles are crucial for a video conference. Fluorescent lights aren't particularly flattering for most people, and they can cast some shadows that don't look great either.

Place your primary source of light behind your camera, this will reduce the risk of unwanted shadows. As for the angle, make sure that your camera is facing you at eye-level. This will help you to make eye-contact with the people in your meeting.

Once you've got everything set up to ensure that you look your best, make sure that you check your look in your webcam. Remember that cameras can sometimes wash people out, which might mean that you need to adjust the brightness to feel confident. If you're concerned about how you look and feeling less confident, try the filters.

Zoom has various filters to help you look your best. The softening filter reduces the sharpness in your image so you can reduce the appearance of dark circles under your eyes. You can also play around with different level of contrast and brightness. Try to avoid going over-the-top with filters unless you know that you're in the right company. Changing your face into a potato, for instance, won't be a good idea in every professional conversation.

Dressing for Success

After you know your face is looking good, it's time to focus on your outfit. There are no particular hard-and-fast rules here. Start by wearing the same kind of clothes you would usually wear if you were having this meeting in person. Sometimes, that will mean choosing smart casual attire. Other times, you might have to go for the full suit and tie combo.

The best way to get started is to just start dressing for work. You might think that you can take advantage of the camera not being able to see below your waist by dressing in comfy pants or trousers. However, remember that there's always a chance that you might need to stand up to get something in the middle of a meeting.

Here are a few quick tips to help you out:

- **Be appropriate:** Choose clothing that's appropriate for the workplace. That means nothing oversized, under-sized, or too revealing. Don't go too casual, even if you're in a reasonably casual setting. You'll need to continue to portray an air of professionalism. A button-up shirt and jeans might be okay, but don't wear your stained graphic t-shirt.

- **Wear colors that flatter you:** This is a fashion tip that works for any situation – including Zoom calls. Since cameras tend to wash you out, avoid any colors that might amplify that effect. Very pale colors that are too close to your natural skin tone can make you look ghostly in a video conference. Additionally, remember that some neutral colors can end up blending into your background.

- **Remember what kind of tech you're using:** If you're using a green screen for a virtual background, it's best to stay away from any green colors in your clothing, as this can leave you looking like a floating head.

- **Be careful with patterns:** Patterns and checks can draw too much attention in some situations. Try to stay away from anything too wild, as it could make it harder for your colleagues and customers to focus on your face. High-contrast patterns are often very distracting, as are bright colors.

- **Don't go over the top with jewelry:** If you decide to wear jewelry, don't go over the top. Small or mid-sized options are okay. However, remember that large, shiny pieces of statement jewelry are often distracting. They can also reflect the light from your computer screen and make it harder for your camera to capture a high-quality image.

Bonus Tips for Looking Your Best

If you're worried about making sure you look your best for your next meeting, consider having some extra tools on-hand for the meeting, just in case. A comb that you can use to brush through your hair when you're on a break between calls is a good idea.

If you're planning on wearing makeup to the meeting, remember that there are pitfalls to wearing too much. Bright lipsticks can look overwhelming on a small screen. Anything that's extremely dark will also look smaller, so be careful to avoid too much dark makeup around the eye.

Check your appearance on the camera before you start each meeting and be ready to make some quick touch-ups if you need to. A great way to ensure that your outfit keeps you motivated and in work mode, is to wear something with a little structure to it. Tailored blouses and shirts encourage us to sit up straighter and stay focused on the task at hand.

Position the screen you're looking at with your video conference so you can look directly at the person, rather than staring somewhere to the side. If your computer monitor is to the left and the camera is too the right this can make it seem like you're not paying attention. Remember your lighting too, shady and dark spaces can make it seem like you're hiding something, or you're just not prepared for the meeting.

Also, remember that there's nothing wrong with wearing clothes and items that make you **feel good.** A bright hue in your clothing can give you a mood boost, so you're more positive in your meeting. A shirt that flatters your physique will give you an extra dose of confidence when you need it most. Take advantage of the power that the right appearance can have.

Chapter 6

Tip 5 for Zoom Meetings: Be Interested

A new psychological phenomenon began to emerge in 2020 with the rise of the video call. Zoom fatigue is the idea that too many regular video meetings can lead to feelings of tiredness and overwhelm. While video platforms have a lot of benefits to offer, they can also be exhausting when you're using them all the time.

Being active on a video call requires a lot of cognitive focus. You need to make sure that you're constantly concentrating on the topic at hand. At the same time, there are a lot of extra things that can grab your attention on the computer screen – including your own image.

Even if you're feeling the onset of Zoom fatigue, it's important to make sure that you're present, interested and in the moment for every meeting. If you feel like you can't focus or give a conversation your full attention, then you're better off re-scheduling the appointment.

If you're responsible for running your Zoom meeting, encourage participants to take part in any way they can. This includes leaving messages in the chat box and linking people to their social media pages or websites if necessary. It's also worth learning how to transition seamlessly from your presentation to your personal video. This will help to engage your audience and keep them focused on both you and your message.

Remember, throughout the course of your Zoom meetings, don't be afraid to take breaks. Being on screen and under scrutiny all day is exhausting. Take some time every so often to turn off and give your brain time to reset. Don't let yourself get stuck in other distractions during these break moments too. For instance, don't switch straight from Zoom to social media.

Organize your time so you can focus completely on your video sessions when you're in them, and don't allow other distractions to get in the way. This also means keeping social media and ringtones from interrupting you when you're mid-conversation.

How to Stay Present in Video Meetings

Now that around 80% of executives believe that video is overtaking audio conferencing as the top form of video communication, no-one can afford to underestimate the power of the visual meeting. It's your job in this conversations to make sure that you're paying attention and showing your full commitment to the conversation.

That starts with looking straight into the camera. If your webcam is situated correctly, it should be easy for you to make eye-contact with the lens. This will translate to better eye-contact with the people on the other end of the conversation. A common mistake is to look at the video feed instead of the camera, but this often means that you end up seeming as though you're not paying attention. You don't want to come across as aloof and unprofessional.

Focusing on the camera, instead of your computer screen or video feed, will also help you to pay attention. While it's okay to look at a presentation that your colleague is sharing or examine a person's body language from time-to-time, those are the only moments when you should be looking away from the camera.

Remember, only 3% of people can actually multitask effectively. The rest of us stop being able to focus when we're not concentrating on one specific thing. Don't allow your email notifications and messages from Slack to distract you. If it helps, close down all of the other tabs and windows that you have open, aside from the one that you're using for your video call.

Whatever you do, don't start typing or shuffling through pages when you're meant to be in a meeting. It's obvious when you're replying to an email rather than taking notes on your conversation. If you need to refer to notes or other information that you've prepared for the meeting, it might make sense to share your screen so everyone can follow along with you. At the very least, you should have the information you need prepared and accessible, so you don't have to waste time searching for it.

To ensure that your colleagues or contacts know that you're genuinely paying attention, remember to regularly speak up and respond to their questions or comments. In moments of silence when another speaker is waiting for a response, acknowledge what is being said.

Manners Make for a Better Meeting

It doesn't cost anything to show some manners and common courtesy during your video conference. That doesn't necessarily mean that you should start a conversation with small-talk about weather. Instead, it means knowing when to concede the floor to other people, and when to step in and speak.

If you're going to be running late to a video conference, don't just jump in at any time and assume it's okay. Send an email or a message letting your contacts know what's going on. When you do join the meeting, apologize for your lateness. Remember that the people who you're talking to have valuable things in their schedule that they need to be doing too. Show them some respect.

It's also better manners to ensure that you always switch your camera on when you're video conferencing with other people. Even if you feel a bit uncomfortable about being on camera, if you're the only person who's audio-only, it makes you look a little suspicious, or less invested in the conversation. Video adds more of a human touch to your meetings, and it shows that you're committed to and focused on what's going on.

If you are on a networking call then people will be more likely to remember you if they see you with a video camera. People take note of the human beings that they can see and interact with on a deeper level. Using video for networking means that you become more memorable and convince your audience that you're professional too.

Remember, one of the best things you can do to show manners in a Zoom meeting is mute your microphone when you're not talking. This will ensure that everyone can give the speaker their full attention, without hearing the white noise in the background of your office.

Chapter 7

Tip 6 for Zoom Meetings: Back Up Your Brand

If you're having a Zoom meeting with a potential investor or partner for your company, then it's essential to showcase the elements of your brand that you want them to remember. There are ways that you can strengthen your brand image both within your Zoom meetings, and outside of them.

For instance, before you even start launching meetings, think about what people are going to see when they research you online. It's not just your company that has a brand to maintain – it's you as an entrepreneur too.

Before starting a Zoom meeting, any good potential partner or investor will want to check you out. The same goes for clients who need to learn more about what you have to offer. With that in mind, start checking your brand image by searching for your name on Google.

What's the first thing to show up when someone looks for you? Do you have any bad reviews from other customers that are cluttering the search results? If so, can you reach out to those people and make amends?

If people try to find you on social media, are they more likely to track down your personal account or your professional account? If your personal account is the first thing to appear, then how does it make you look as a professional? People won't just give you a break for acting poorly on your personal social media. If you're concerned about your personal accounts' influence on your business image, make the accounts private.

Manage Your Social Media Presence

After you've searched for your name on Google, take the time to go through your social media accounts, and make updates from a professional perspective. If you don't have dedicated accounts on LinkedIn, Twitter, and Facebook already, now could be the time to make them.

Don't make the assumption that staying off social media is better than having a bad account. If people can't find information out about you online, then they're going to feel more suspicious. Not having a social presence is sometimes worse than having a poorly optimized one because it suggests that you have something to hide.

Since your LinkedIn profile will be the most reputable environment that your audience looks at before connecting with you on Zoom, start your strategy there:

- **Focus on industry skills:** Recruiters, partners, and clients will often search for keywords that related to what they need. It might be worth including terms in your LinkedIn profile that will help people to find you and understand what you focus on as a professional.

- **Showcase your accomplishments:** Make it easy for the people who find you to see what makes you special. Highlight the work you've already done in your industry and quantify your results. Don't just say that you've built one of the best shoe companies in the world. Explain why you think that way. How many customers do you have, what are your figures like?

- **Complete your profile:** An incomplete profile on LinkedIn will either look lazy or suspicious. Neither option is good for the brand you're trying to create. Focus on filling your LinkedIn profile with all the information you can think of. Share your educational background, your entrepreneurial accomplishments, and anything else people might want to know.

- **Use a professional photo:** LinkedIn users that include a professional headshot on their account get around 14 times more views than people without a high-quality image. Upload a photo that looks great and makes you seem approachable. If you can't find a professional picture, stick with your logo instead until you get one, but get one.

- **Show that you're a thought leader:** You can share content and links on LinkedIn – and this is a great way to demonstrate your knowledge in your space. Your thought-leadership pieces on LinkedIn might connect back to videos and podcasts that you've published elsewhere on the web. The more you can demonstrate your expertise, the better.

- **Place a branded banner into the profile section** – Missing this is like having a shop with no sign above it. Think about how you can instantly grab attention with all the components of your LinkedIn profile and online presence.

Remember to be consistent with your personal brand too. If you highlight specific characteristics about yourself on LinkedIn, then continue to highlight them on your Twitter and Facebook pages too. If you can, invest in a personal website or portfolio to further establish your presence in your industry. Although some people will give you the benefit of the doubt, if you don't have a website, most investors and potential partners will expect you to spend money on a site that showcases your work.

Bring Your Brand to the Video Meeting

You can also bring elements of your personal brand into the video meeting. For instance, instead of just having a random virtual background on Zoom, upload a high-quality image that includes a version of your logo. This will make you look more professional, while also reminding your audience of one of your most valuable brand assets.

If you're not using a virtual background, you might have a picture of your logo behind you sitting on the desk, or a picture of you with the rest of your team, to highlight your professional stance. Drink your coffee out of a mug that has your logo plastered onto it. Wear a T-shirt with your own business name written on it (professionally).

This kind of self-promotion in a business meeting can sometimes come across as tacky if it's not done well. However, adding elements of your brand to your video meeting can also make you more memorable if you do it in a clean and sophisticated way.

Chapter 8

Tip 7: Design Clean Presentations

Zoom isn't just for having conversations with your partners, colleagues, and shareholders. Video conferencing tools are becoming a valuable part of the business development landscape. It's helpful for all kinds of meetings and interactions, whether you're networking or developing new opportunities with team members. Unfortunately, a lot of the people who are shifting to Zoom and similar tools for Business Development aren't taking full advantage of its benefits.

You can also use the video platform to present your ideas to other people. If you want to pitch a new idea to your investors, Networking and conferencing or onboard some new freelance workers, then Zoom can help you to do just that. The key to success is knowing how to prepare for a successful presentation.

If you have the opportunity to showcase something important through video, then you need to make sure that you do so clearly, and effectively. Start by thinking carefully about the demographics of the people on the call. You need your presentation to appeal to the right audience.

Getting Ready for a Presentation

Getting ready for your Zoom presentation is just as important as actually giving it. Just like preparing for a meeting, you'll need to do some initial tech checks first. Make sure that your microphone and camera are working as they should be. It's also worth checking:

- **Your internet connection:** Using an ethernet cable instead of your Wi-Fi is probably a good idea here. Nobody wants to listen to a presenter with laggy video and audio. A high-speed internet connection will ensure that your presentation goes smoothly.

- **Your power supply:** If you're using a laptop, make sure it's plugged in before you start your presentation. You don't want it to die when you're in the middle of talking.

- **Your environment:** Make sure you have the right lighting and environment for your presentation. If you're going to be showing off a whiteboard, make sure that your audience can see it properly. Don't stand in front of the presentation.

- **Your distractions:** Just as you don't want members of your family wandering in on the background of your meeting and distracting you, you don't want any social media prompts jumping up on your presentation and making you look less professional. Turn notifications off to reduce distractions and minimize embarrassments.

- **Your audio:** From an audio perspective you'll need to check a few things. First, make sure that your audience can hear you clearly with the microphone you're using. Next, ensure that all other audio distractions are switched off (including your mobile).

If you want your presentation to be particularly good, then you could consider presenting your slide deck with a second monitor. This allows you to use a new feature called "Presenter view". Essentially, this allows you to privately display your current slide, your next slide, and your speaker notes. Plus, you get a clock and timer to help you stay on track.

This is a useful feature to have in an online environment because it stops you from getting distracted or having awkward transitions when you forget what the next slide is about.

Remember to get rid of anything that might distract people in the presentation too, such as the chimes that happen when someone joins a conversation. You can even ask people to mute their audio when they're coming into the presentation to stop them from disrupting the flow of your speech if you've already gotten started.

Although a virtual background can be a good idea in some scenarios, in others it's just distracting. If you use your hands a lot, then the ghost effect caused by a virtual background is often very off-putting. A better option might be to simply tidy up your real-world background.

It may be worth turning your own video and audio off before you're ready to start the presentation too. This will give you a chance to ensure that everything is set up properly before jumping in. To mute your audio and turn off video when joining a meeting:

- Go to your Zoom desktop applications
- Click on **Preferences** then **Video**
- Check the box for **turn off my video when joining a meeting**
- (Optional) check the box to "touch up my appearance) to airbrush your skin
- Go into the **audio menu** and click **mute microphone when joining a meeting**
- Apply your changes

Improving the Quality of your Presentation

If you're presenting a slide deck or sharing information in your video presentation, make sure that people don't get distracted by other streams. During the meeting, hover over your own video window to show the three dot menu. Click on that then **spotlight video** to keep the focus of all attendees on your stream.

Here are some other tips to improve the quality of your presentation too:

- **Keep it simple:** The best presentations are simple and easy to follow. Don't bog your audience down with too much information to keep track of. Stick to simple presentations that are easy to follow. Slides that are straight to the point are helpful here. Remember, you can always send extra information through Zoom chat or email later.

- **Make the presentation easy to follow:** To ensure that your attendees don't get too distracted, use a plain background for your slides, and focus on short snippets of text. Centering your text in the slides is often helpful too. A large bold font will improve clarity.

- **Put your notes in the right place:** If you're using notes to keep you on track during the presentation, make sure that you place those notes in the right space. It often looks unprofessional if your eyes are constantly moving away from the camera. Sticking post-its below and next to your webcam will ensure you seem as focused as possible.

- **Rehearse before the presentation:** Don't walk into a presentation unprepared. This is particularly important when you're on Zoom, and you don't have the people around you to give you immediate feedback and boost your confidence. If it helps, consider running a mock presentation with a friend first. Just don't share any sensitive information.

- **Keep your attendees engaged:** Make sure that people don't get bored and switch off during your presentation. Ask questions to get people involved and make sure they're engaging with your content. Request feedback during the conversation and answer any queries that your attendees might have. You can even give your Zoom attendees the option to raise their hand when they have something important to share. Which is a feature of the program both through the supportive on option on the tool bar "disappears after a few seconds "and the "Blue" raise a hand option at the bottom of the chat box which remains until cancelled off by the host.

CHAPTER 9

Tip 8: Stand Out

If you were meeting with a prospective partner, customer, or investor in person, you'd want to make an impact. First impressions count – and you still make a significant impression over video.

Remember that around 55% of the assumptions we make about someone come from what we see. In a face-to-face setting, the way you hold yourself, the meeting room, and your general appearance will all make a difference. In a video conference, the people you speak to are still going to be considering those factors.

Start by thinking about how you appear in the chat. Are you active and engaged in the conversation? If you've set your environment up correctly and dressed correctly then you should make a good initial visual impression. However, people will also be looking at the way that you respond to the conversation, and how much you're taking part in the meeting. Ensure you're an avid contributor by:

- **Nodding and agreeing:** You don't have to verbally respond to everything that someone says, but nodding is a good way to demonstrate that you're taking the message in, and you agree with it.

- **Use reactions and animations:** There are reactions available on Zoom meetings which allow you to send a thumbs up, or a picture of some hands clapping to convey your mood at any given moment. These reactions may not be appropriate all the time, but they're great in a slightly less formal setting.

- **Responding (politely):** When the person in your meeting finishes talking, you can respond to what they've said in a meaningful way. Reference keywords that they've mentioned when giving your response and give your own opinions. Don't just nod and say "ok".

If you can't respond verbally straight away, you could always send a message in the chat. This will act as a note to remind you of something you wanted to say when there is time to chat.

Planning to Participate in a Zoom Meeting

Making sure you respond correctly to everything that people are saying in a Zoom meeting sometimes means that you need to organize your thoughts and ideas before the conversation starts. Zoom fatigue is an increasingly common problem, with people spending hours on video calls every day.

You need to be concise and focused on what you want to say and share. As you might do for an in-person meeting, try to plan carefully, and anticipate what you're going to say and share. Expect that the people speaking to you will probably have a shorter attention span or become more easily distracted in these meetings too.

Don't just make the conversation all about you – though. Remember that you need to stand out in any environment by actually engaging and responding to the things that other people are saying. The only way to respond in a valuable way is to actually pay attention to what people are saying. Your attentiveness won't go unappreciated.

Remember that body language makes a big difference in video conferencing too. You're on display, which means that like in a face-to-face meeting, you shouldn't be slumped in your seat, or looking away from the camera. Show your audience that they have your full attention by sitting up straight and working on your posture.

Quick Ways to Stand Out in a Zoom Meeting

The good news for those who really want to ramp up their appearance in a Zoom meeting, is that there are a lot of quick ways to stand out. For instance, you can add a virtual background to your Zoom meeting that shows off your brand logo.

It's also a good idea to spend some time checking out all of the various features available on Zoom that can make you look more professional. For instance, using the Zoom calendar integration so you can schedule follow-up meetings instantly with people you need to talk to again is a great step.

You can also set up Zoom waiting rooms, so when you're preparing any demonstrations, your participants have a valuable place where they can wait. Remember you can always touch up your face too to get rid of anything that might make you look a little tired or strained. Touching up your appearance makes you look fresh-faced and ready to go for any conversation.

When you're ready to start speaking, make sure you have your notes somewhere close by, so you don't have to turn your attention away from the camera and screen. Use assertive language that highlights your knowledge in your industry, and make sure that you narrow messages down to key points that you want your audience to take away.

If it helps, you can send the people from your meeting some notes following the conversation that reminds them exactly what you spoke about, and what's going to happen next.

Chapter 10

Tip 9: Master Your Background

We've mentioned the importance of backgrounds on Zoom a few times so far. That's because virtual backgrounds are some of the most valuable tools available for Zoom users. When you're trying to present a professional brand to the people you're meeting with, virtual backgrounds allow you to block out unwanted distractions and mess that might pull attention away from you in a conversation.

You don't always have to stick with virtual backgrounds to succeed on Zoom. If you've got a decent looking space that you can take your meetings in, that's fine too. A white background is usually a good choice, as long as it isn't too bright, or worn with white shirts that drain your color.

You should always make sure that anything visible in your meeting isn't detracting from your background though. Although the people you're speaking to might expect you to have things like washing hung up in the background if you're at home, they also expect you to have enough fore-thought to get rid of those things before you start a meeting.

How to Use Virtual Backgrounds Correctly

Customized virtual backgrounds can be a lifesaver when you're having conversations over video conference. When the work-from-home revolution began to really take hold in 2020, there were countless stories of people with unexpected things in the background of their video calls. We've had family members walking into the background of meetings, dogs running around, and even loved ones walking into the screen barely dressed. Unless you are running an online networking solution for Pet Owners (like I do).

A virtual background can help to reduce the risk of these problems, although we do recommend making sure that you have a room for your office with a lock on the door just in case. Being able to lock the outside world away during online meetings will save you a lot of headaches. It also means that you can enjoy a better quality of work-life balance, because you can close and lock your office door when you're done working for the day.

The first thing you need to know about Zoom virtual backgrounds is that you do get a handful already included with the app. You can also add images from your photos file on your computer to customize how you look on-screen. For instance, you might have a picture of a professional office that you can use when you're at home to convey that you're just as ready to work in any environment.

One thing you should never do is upload a background that's just funny or different. This probably isn't the time to portray yourself as coming to a meeting from the Starship Enterprise. You want to appear professional and focused, which means making sure that you choose a background representative of your brand, unless you think you're in the right environment to do something unique.

The caveat here is you can have a fun or silly background if you are in a relaxed environment and trying to bond with a funny background that doesn't impact your personal brand negatively .e.g. a quiz night with friends.

When in doubt, stick to something neutral, such as a white space or a plain background that makes it look as though you're in an office. If you take pictures for your own Zoom backgrounds, make sure that you capture the images in the highest resolution possible. Other tips to keep in mind:

- Don't try to be witty and interesting: Some people will take the ideas that you consider to be fun and smart the wrong way. It's best not to guess at someone's sense of humor

- Avoid intricate and complicated patterns: These strain the eye and make it harder for your meeting participants to focus on you.

- Stay away from neon colors: Avoid bright colors that drain you or make you more difficult to look at for other meeting attendees.

Tips for Better Conferencing with Virtual Backgrounds

Keep in mind that not **every** computer will necessarily support Zoom backgrounds to their full potential. You'll need at least OS 10.15 on a Mac, and on a PC, you're going to need Windows 10 at a minimum. To support Zoom, you'd ideally need 8 MB of RAM too.

Once you've checked to ensure that your computer is capable of running virtual backgrounds, you can upgrade the quality of your meetings by:

- Getting the lighting right: Make sure that you choose something that lights you from the front, instead of the back, this will be a lot more flattering, and reduce the risk of shadows around your eyes and neck. If you're using a laptop, make sure it's eye height too.

- Using solid colors: Choose a solid color with your virtual background if possible. This will help to reduce the risk of distractions for your viewers and make you look more professional overall. Pick colors that are going to complement your skin tone.

- Staying cool: Get rid of any bright lights behind you like open windows or sunlight. This might mean closing a blind or adjusting the position of your desk. A neutral background will give you a more balanced image.

To boost the overall quality of your meeting image, consider placing a white background behind your head and shoulders. It could be a curtain, a white screen, or even a large piece of paper if that's all you have. This will do wonders to reduce that horrible black outline that can surround your head in some cases. If your virtual background displays a lot of brightness, the white bit above your head and shoulders will make everything look more realistic.

If you can get a green screen then this might help to add a better level of clarity to your virtual meetings. Although you can use a virtual background without having to worry about things like green screens, Zoom does recommend using one if you can.

Other Tips for Backgrounds

One great option to consider if you want to ensure that your people feel more immersed in the meeting experience, is to use the **Immersive Scenes** feature of Zoom. If you're familiar with Microsoft Team's Together mode, this is a very similar solution that allows you to embed users in a virtual environment, so it seems like you're all sitting together in the same room.

Virtual environments like this can help to reduce feelings of meeting fatigue and make your contacts feel more "engaged" in the space. Rather than seeing a grid full of video feeds on your screen, you're seeing a landscape that makes it look like you're all sitting together in the real world.

If you don't want to mess around with virtual environments, but you do want to ensure that your video quality is amazing, then make sure you check your video settings before the conversation begins.

Go into your Zoom application, then click on the **gear icon** in the upper right when you log on. Another option is to go to the lower-left corner of the Zoom screen and click on the **up arrow** next to the video and choose **video settings.** From the next screen, you can begin experimenting with and perfecting your videos according to your needs.

Usually, you'll find that your settings are all default, for things like brightness, contrast and beyond. While this is fine for some people, it won't be the right option for everyone – particularly depending on your camera. You might decide that you want to enable high definition video, for instance. You can also decide whether you want to add participant names to the videos you see.

Chapter 11

Tip 10: Follow up On Meetings

A successful video conference is only the first step on a long road to success.

In today's digital world, where people have dozens of video conversations every day, you can't rely on your combination of exceptional wit and great Zoom skills to make you stand out from the crowd. It's up to you to ensure that you're connecting with the people you meet in an appropriate manner.

Within 48 hours of any video conversation, follow up on the video discussion via email. This is a great opportunity to reflect on the things you discussed in the conversation and address the topics you would like to look at further. You could even use an app that automatically transcribes your meeting on Zoom so you can send your contact a copy of the discussion.

In the email follow-up, call the recipient by their full name, and thank them specifically for what they brought to the meeting. Don't just say "thanks for your time". Reference your specific conversation. This could mean that you say you were thrilled to hear they were interested in using your service, or that you're excited to get started on their new project.

Reference a specific detail of the conversation from your meeting. The more detailed and specific you are with your message, the more authentic it feels. Remember, 77% of people love hearing a genuine thank you, but they won't appreciate it as much if your thank you seems fake or forced.

Follow Up on Group Meetings Too

Remember, while following up on your one-to-one meetings is important, making sure that you make the right impression in a group networking session is important too. While you're in the group chat, do everything you can to stand out – without looking obnoxious. Using the chat, you should be able to comment when the right time arises and get involved with things like hand-raising and emoji responses from time to time too.

Remember, always use your video – don't be the person that hides themselves because you're feeling nervous, as this will mean you're just not memorable. When you're done with the meeting, check out the profiles of the people you met on LinkedIn afterwards, and add them to your network if you can, or send them a message.

During the meeting, make notes (subtly) on who you would like to follow up with, and when following up might be a good idea. Arrange another chat in the future with a focus on getting to know each other a little better. If, during the meeting, you agree to connect or help someone out with something later – do it.

In a world where people just don't understand follow up most of the time, this is your chance to stand out and boost your network. You can follow up and connect in no time!

What to Remember When Following Up

When you're following up on a video conversation, it's important to keep your goal in mind. What was the purpose of having the conversation with this person in the first place? What do you need to do next now that you've had that first conversation? Are you going to offer your assistance with something they're working on, or do you need to make an introduction to the rest of your team? Your follow up messages should always be:

- **Specific:** As mentioned above, don't use the same bottled and generic response for everyone. You should be able to mention specifically what you liked about the meeting and refer to the topic you discussed in detail. Don't use the same response for everyone.

- **Genuine:** People make connections with other people. Make sure that you create a genuine connection with your audience by showing your unique personality. Use the same formal or informal language you used in the meeting in the email and remember to include your logo or something specific to highlight your brand.

- **Timely:** Remember your customers, colleagues and other contacts are having tons of video calls all of the time these days. You don't want to give them too much time to forget about you. Ideally you should follow up with a conversation in 24 hours via email. Within a week, reconnect on social media, and make sure that you have your second video call scheduled too. Usually, it's best to plan a second video or real world conversation within a month of the first.

- **Valuable:** Give the person you've been talking to a reason to continue with the conversation. The benefit is obvious for you. Whether you're getting a new employee, colleague, or investor out of the deal, your business grows. So what does coming to the next meeting do for the other person? Remember "WIIFM" – What's in it for me? (or in this case, for them)

- **Engaging:** Follow up messages should quickly grab your audience's attention and give them a variety of ways to get in touch. Don't just let your contact know that you want to meet for another zoom conversation a week from now. Tell them what you're going to be discussing, address some key takeaways, and give them alternative options too. Zoom might not be the ideal tool for everyone you interact with.

Remember, to make life a little easier for yourself, you can set up templates for the kind of follow-up messages you might send and when you need to send them. The key is to make sure that you personalize each template every time you need to connect with someone.

Don't forget to do whatever you can to make keeping in touch easier for your contact too. Ideally, you should have a way to ask them whether they want to add a new meeting to their calendar, so they get a reminder when a new conversation is due. You could also add a link to your Zoom meeting room within your email signature, reminding the contact where they need to go.

Make Each Connection Better than the Last

As you continue to invest in valuable relationships with professionals through video conferencing and other tools, make sure that you pay attention to any opportunities to learn that might arise along the way. For instance, you might be able to ask your contacts what they liked about your latest video presentation or ask for suggestions on what they might like to do differently in your next meeting.

Don't just send people an agenda telling them exactly what you're going to cover the next time you meet. Instead, provide your contacts with an opportunity to make their own suggestions and adapt the schedule to suit their needs. Perhaps you'll learn that some of your contacts respond to your messages better with visuals like PowerPoints and slideshows.

Maybe you'll discover that although some of your contacts do enjoy using video conferencing tools, they prefer to use a service other than Zoom, such as something that has special security features and data storage solutions that are crucial to their industry. Be ready to adapt to the people in your meeting and their unique needs.

Chapter 12

Tip 10$^{1/2}$: Go Back Again

Networking is one of the most challenging parts of building a successful business (checkout KISS the Rebel on amazon). Most people don't totally understand how to build an effective network for their organization. This means that they frequently see "getting the meeting scheduled" as the end goal. Tracking down the ideal connection, convincing them to have a conversation with you, and displaying your brand properly through that discussion is complicated.

As you may have seen from all the points we've made above, there's a lot to consider during the various stages of your video conversation, from making sure that you know how to use the technology to the best of your ability, to ensuring that you have the right presentations in place.

However, the networking conversation isn't the end of the journey for today's business leaders – it's the beginning. If you want a connection with another individual or business to go in the right direction, then you need to think of building that relationship as the same as making any other connection.

If you make a new friend in your everyday life, that friendship won't lead to much if you never follow up with your new acquaintance or check in on how they're doing. The more time you spend with another person, the more you grow to trust and understand them. Over time, you and your contact contribute to each other's lives, and real relationships are born.

Networking follows the same pattern. In the world of business, networking isn't just about selling your goods and services to other leaders face-to-face. You might not even go into your first networking conversation with any intention to sell at first – unless the person you're speaking to asks whether you can suggest a solution to their problem.

Networking is more about building genuine and valuable relationships that you and your contact can leverage for mutually beneficial outcomes.

Maintain Your Valuable Relationships

If your relationship with a new investor, colleague, or innovator starts with a Zoom meeting, it doesn't necessarily need to end there too. You should look at each relationship with individual people as a unique opportunity. With one person, you might find that regular video meetings are the best way to generate a sense of connection and camaraderie.

On the other hand, for another contact, you might get better results from regular face-to-face interactions, coffee breaks, and discussions where you can actually interact and brainstorm ideas together.

Remember that following up doesn't necessarily have to be about pushing a sale either. You can also just reach out to develop a more significant emotional connection with someone.

Asking to meet up for coffee with a colleague is a great way to generate a sense of emotional bond between you and that person. The strength of their connection with you as a person will eventually influence how they think and feel about your company – particularly if you're the face of the brand. When investors feel they can trust and understand entrepreneurs, that's when they're more likely to get involved and start spending money.

You could also use each follow-up session as a chance to further demonstrate the value that you can bring to the relationship in question. Links to curated blog posts that are interesting to your chosen contact, or references to new events could have a massive influence on how frequently your new connection comes to you for guidance or information.

If you and your contact find that regular meetings contribute to a mutually beneficial relationship for both of you, then you could always schedule for regular meetings with a digital calendar app. Many of these tools already integrate seamlessly with things like Zoom. They can link with your messaging apps and email inbox to remind you

Embracing the Future of Networking

Zoom is just one tool in a huge selection of valuable resources that today's business leaders can use. Video conferencing tools like Zoom provide us with a way to connect with our colleagues and other professionals in an important way. When face-to-face networking isn't possible, video conferencing can be an excellent way to maintain and enhance human relationships.

Of course, that doesn't mean that you should be relying on video for your networking efforts exclusively. There's still plenty to be said for tactics like sending messages over social media, getting involved with group forums, and even writing the occasional email.

The key to success is figuring out how each contact responds to the different methods of interaction you have to offer and being open to new ideas. Ten years ago, regular remote working, and video meetings might have been an impossible thought. Another decade in the future, we could be interacting with holograms, virtual reality, and augmented reality without batting an eyelid.

Take advantage of Zoom but be ready to embrace all of the new and evolving tools for human connectivity that are emerging in the landscape today. The digital transformation of networking opportunities is accelerating at an incredible pace. Who knows what the future might bring?

www.ingramcontent.com/pod-product-compliance
Lightning Source LLC
Chambersburg PA
CBHW021023180526
45163CB00005B/2090